AF808306

LEMONLearn
Learners Today. Leaders Tomorrow.

TODDLER LEAF MATH WORKBOOKS 1, 2, 3

1 2 3 4 5

6 7 8 9 10

LEMON Learn

Learners Today. Leaders Tomorrow.

HOW TO USE THIS BOOK

Encourage your early learner to complete each booklet at their own pace.

1 OPENING PAGE

Each booklet has an opening page. Before starting each section, this page will show you what booklet and skill your child will be focusing on.

2 STEPS 1, 2, 3

Steps 1, 2, 3 are on each opening page to guide you while working with your child.

3 NAME, DATE, TIME

Instruct your learner to write their name, date, and start time before beginning their work. When their work is complete, have them write their finish time. **If the student can not write yet, write it for them.**

HOW TO USE THIS BOOK

Get your student talking! Ask them about the images they see and connections between the workbook and their lives.

4 DIRECTIONS

Have students read the directions aloud on each page, if possible. **If the student can not read yet, read the directions aloud to them.**

LEGEND 5

- Page Number
- Subject
- Booklet Number
- Skill

6 LEAF TIPS

Leaf Tips give parents detailed instructions and assistance for helping with each section.

I DID IT! 7

Guide students to draw a check mark in the box. This positively enforces activity completion. Have students say, "I did it!" to get excited about learning.

HOW TO USE THIS BOOK

Students can practice the skills they learn with the math games!

8 MATH GAMES

At the end of **Booklets 1 – 3**, you will find a math games page that helps introduce numbers during everyday activities with your learner. This activity requires crayons.

POSTERS 9

At the end of **Booklet 3**, you'll find an encouraging poster to celebrate your child's learning. Cut it out and display it on the wall or let your child share it with friends.

CONCEPTS

Booklets 1 - 3

Stable Order Principle

Understanding that the counting sequence stays consistent. Its always 1, 2, 3, 4, 5, 6, 7 ect. NOT 1, 2, 4, 3 ect.

Order Irrelevance Principle

Understanding that the counting of objects can begin with any object and the total will stay the same.

One-to-One Correspondence

Understanding that each object being counted must be given one count only. Children tag each item being counted by pointing and saying the numerical name aloud.

> "Learners should work as independently as possible. Guide them to approach each page with a try-again mindset. Every challenge provides an opportunity to learn and grow to achieve success."

FLEXIBILE LEARNING

At Lemon Learn, we understand that every child develops at their own pace, and it's never too early to introduce math—even to children who aren't speaking yet. They can begin this workbook by simply listening to parents recite the oral Booklets 1 - 3 and enjoying the math games that follow. Once your child starts speaking, you can begin again from Booklet 1 to reinforce their learning from the start.

Booklet 1

Counting up to 5

1. Preview the pages of Booklet 1 with your student before starting.
2. Help your student write their name, date, and start / finish time.
3. Work together with your student for 15-30 minutes.

Name: ____________________

Date: ____________________

Time: ____ : ____ **to** ____ : ____

1a

LEMONLearn

Leaf Math

Booklet 1

Counting 1

(up to 5)

Count the pictures aloud while pointing to each one.

LEAF TIPS

- Read the directions aloud on each page.
- Demonstrate how to count the pictures and dots (●) on each page: Using the index finger, point to each one while counting aloud.
- After each activity, check the box at the bottom of each page. Have students positively reinforce activity completion by saying, "I did it!"

I did it!

☐

Count the dots aloud while pointing to each one.

LEAF TIPS

- Students may point to the pictures and dots (●) in any order: left-to-right, right-to-left, top-to-bottom, and bottom-to-top.
- Each dot should be counted only once no matter what order they were counted in.

I did it!

☐

2 a

LEMONLearn

Leaf Math
Booklet 1
Counting 1
(up to 5)

Count the pictures aloud while pointing to each one.

I did it!

☐

2 b

LEMONLearn

Leaf Math

Booklet 1

Counting 1

(up to 5)

Count the dots aloud while pointing to each one.

I did it!

LEMONLearn

Leaf Math
Booklet 1
Counting 1
(up to 5)

Count the pictures aloud while pointing to each one.

I did it!

☐

Count the dots aloud while pointing to each one.

I did it!

☐

4 a

LEMON Learn

Leaf Math
Booklet 1
Counting 1
(up to 5)

Count the pictures aloud while pointing to each one.

I did it!

Count the dots aloud while pointing to each one.

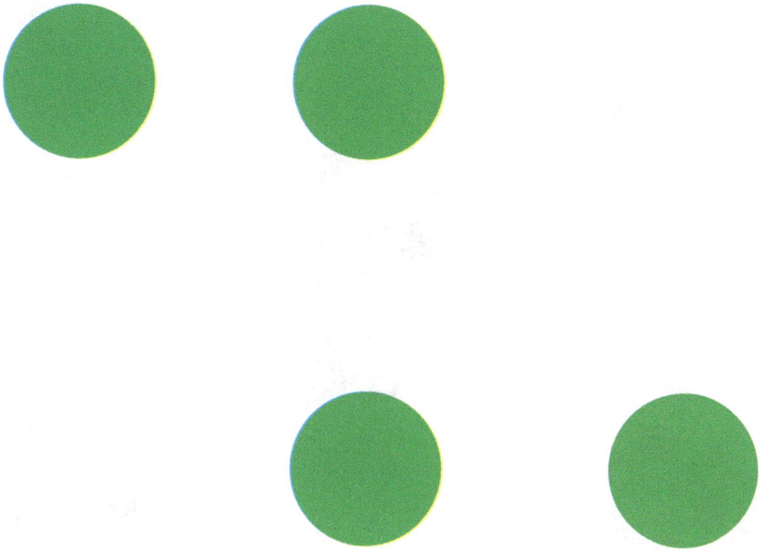

I did it!

☐

LEMONLearn

Leaf Math
Booklet 1
Counting 1
(up to 5)

Count the pictures aloud while pointing to each one.

I did it!

☐

Count the dots aloud while pointing to each one.

I did it!

☐

6 a

LEMON Learn

Leaf Math

Booklet 1

Counting 1

(up to 5)

Count the pictures aloud while pointing to each one.

I did it!

6 b

LEMONLearn

Leaf Math
Booklet 1
Counting 1
(up to 5)

Count the dots aloud while pointing to each one.

I did it!

LEMONLearn

Count the pictures aloud while pointing to each one.

I did it!

☐

7 b

Leaf Math
Booklet 1
Counting 1
(up to 5)

Count the dots aloud while pointing to each one.

I did it!

LEMONLearn

Count the pictures aloud while pointing to each one.

I did it!

Count the dots aloud while pointing to each one.

I did it!

LEMONLearn

Leaf Math
Booklet 1
Counting 1
(up to 5)

Count the pictures aloud while pointing to each one.

I did it!

Leaf Math

Booklet 1
Counting 1
(up to 5)

Count the dots aloud while pointing to each one.

I did it!

☐

Count the pictures aloud while pointing to each one.

I did it!

☐

LEMON Learn

Count the dots aloud while pointing to each one.

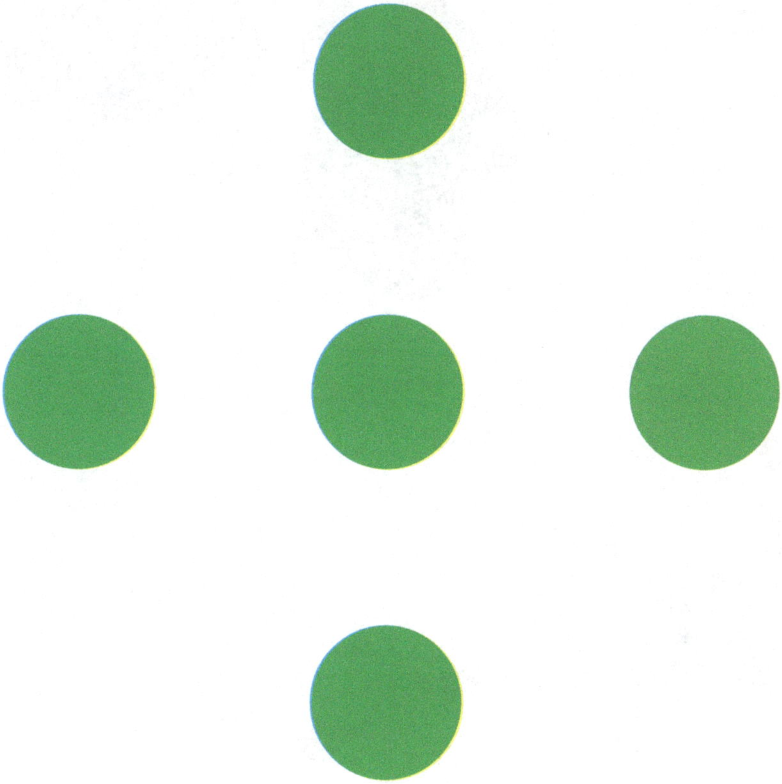

I did it!

☐

PRACTICE NUMBERS EVERYDAY

Learn beyond the classroom during everyday activities!

MATH WALK

Take a math walk together. As you walk, play a game where you and your child need to pick "1 flower," collect "2 leaves," take "3 big steps," find "4 pebbles" and touch "5 trees." Count aloud as you do each task.

Booklet 2

Counting up to 5

1

Preview the pages of Booklet 2 with your student before starting.

2

Help your student write their name, date, and start / finish time.

3

Work together with your student for 15-30 minutes.

Name: ______________________

Date: ______________________

Time: ____ : ____ **to** ____ : ____

LEMONLearn

Count the pictures aloud while pointing to each one.

LEAF TIPS

- Read the directions aloud on each page.
- Demonstrate how to count the pictures and dots (●) on each page: Using the index finger, point to each one while counting aloud.
- After each activity, check the box at the bottom of each page. Have students positively reinforce activity completion by saying, "I did it!"

I did it!

☐

11b

LEMONLearn

Leaf Math

Booklet 2

Counting 2

(up to 5)

Count the dots aloud while pointing to each one.

LEAF TIPS

- Students may point to the pictures and dots (●) in any order: left-to-right, right-to-left, top-to-bottom, and bottom-to-top.
- Each dot should be counted only once no matter what order they were counted in.

I did it!

☐

Count the pictures aloud while pointing to each one.

I did it!

LEMONLearn

Leaf Math
Booklet 2
Counting 2
(up to 5)

Count the dots aloud while pointing to each one.

I did it!

☐

13a

LEMONLearn

Leaf Math
Booklet 2
Counting 2
(up to 5)

Count the pictures aloud while pointing to each one.

I did it!

☐

Count the dots aloud while pointing to each one.

I did it!

☐

Count the pictures aloud while pointing to each one.

I did it!

Count the dots aloud while pointing to each one.

I did it!

☐

Count the pictures aloud while pointing to each one.

I did it!

15b

LEMONLearn

Leaf Math

Booklet 2

Counting 2

(up to 5)

Count the dots aloud while pointing to each one.

I did it!

LEMONLearn

Count the pictures aloud while pointing to each one.

I did it!

Count the dots aloud while pointing to each one.

I did it!

☐

Count the pictures aloud while pointing to each one.

I did it!

☐

Count the dots aloud while pointing to each one.

I did it!

LEMONLearn

Leaf Math
Booklet 2
Counting 2
(up to 5)

Count the pictures aloud while pointing to each one.

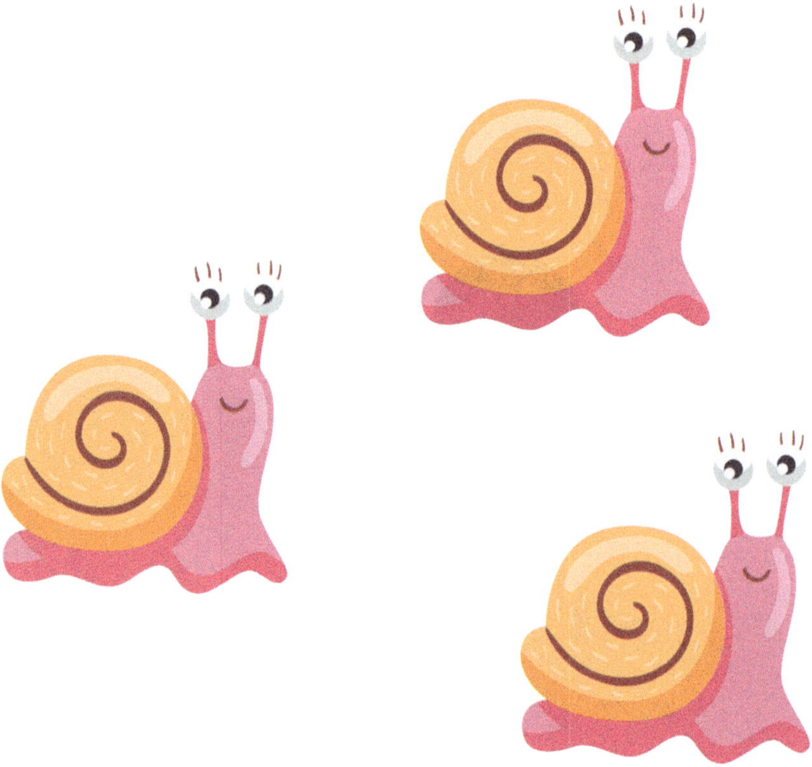

I did it!

☐

LEMONLearn

Count the dots aloud while pointing to each one.

I did it!

Count the pictures aloud while pointing to each one.

I did it!

LEMONLearn

Count the dots aloud while pointing to each one.

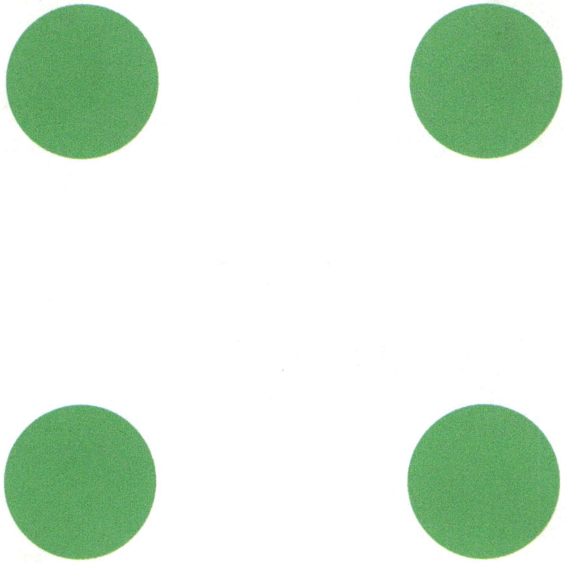

I did it!

☐

20a

LEMONLearn

Leaf Math
Booklet 2
Counting 2
(up to 5)

Count the pictures aloud while pointing to each one.

I did it!

Count the dots aloud while pointing to each one.

I did it!

☐

PRACTICE NUMBERS EVERYDAY

Learn beyond the classroom during everyday activities!

LOVE COUNTS

Before bedtime, ask if your child wants "2 kisses" or "3 kisses." Count aloud as you give each kiss. When waking up, ask if your child wants "1 hug" or "2 hugs." Count aloud as you give each hug.

Booklet 3

Counting up to 5

1 Preview the pages of Booklet 3 with your student before starting.

2 Help your student write their name, date, and start / finish time.

3 Work together with your student for 15-30 minutes.

Name: ______________________________

Date: ______________________________

Time: _____ : _____ to _____ : _____

LEMONLearn

Count the pictures aloud while pointing to each one.

LEAF TIPS

- Read the directions aloud on each page.
- Demonstrate how to count the pictures and dots (●) on each page: Using the index finger, point to each one while counting aloud.
- After each activity, check the box at the bottom of each page. Have students positively reinforce activity completion by saying, "I did it!"

I did it!

LEMONLearn

Count the dots aloud while pointing to each one.

LEAF TIPS

- Students may point to the pictures and dots (●) in any order: left-to-right, right-to-left, top-to-bottom, and bottom-to-top.
- Each dot should be counted only once no matter what order they were counted in.

I did it!

Count the pictures aloud while pointing to each one.

I did it!

☐

Count the dots aloud while pointing to each one.

I did it!

☐

LEMONLearn

Count the pictures aloud while pointing to each one.

I did it!

☐

Count the dots aloud while pointing to each one.

I did it!

LEMONLearn

Leaf Math
Booklet 3
Counting 3
(up to 5)

Count the pictures aloud while pointing to each one.

I did it!

☐

LEMONLearn

Count the dots aloud while pointing to each one.

I did it!

Count the pictures aloud while pointing to each one.

I did it!

LEMONLearn

Count the dots aloud while pointing to each one.

I did it!

26a

LEMONLearn

Leaf Math
Booklet 3
Counting 3
(up to 5)

Count the pictures aloud while pointing to each one.

I did it!

LEMONLearn

Count the dots aloud while pointing to each one.

I did it!

Count the pictures aloud while pointing to each one.

I did it!

☐

LEMONLearn

Leaf Math
Booklet 3
Counting 3
(up to 5)

Count the dots aloud while pointing to each one.

I did it!

☐

Count the pictures aloud while pointing to each one.

I did it!

Count the dots aloud while pointing to each one.

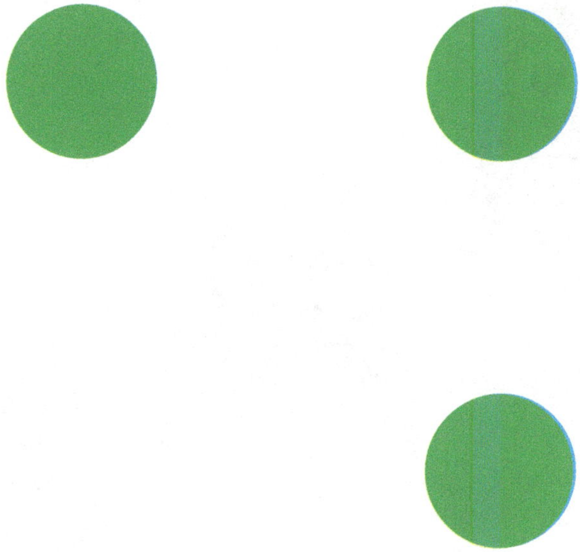

I did it!

☐

Count the pictures aloud while pointing to each one.

I did it!

Count the dots aloud while pointing to each one.

I did it!

☐

LEMONLearn

Leaf Math
Booklet 3
Counting 3
(up to 5)

Count the pictures aloud while pointing to each one.

I did it!

LEMONLearn

Leaf Math
Booklet 3
Counting 3
(up to 5)

Count the dots aloud while pointing to each one.

I did it!

PRACTICE NUMBERS EVERYDAY

Learn beyond the classroom during everyday activities!

COUNT & CLEAN

After playtime, count toys during cleanup. For example, ask your child to put away "1 teddy bear," stack "2 books," organize "3 cars," gather "4 blocks," arrange "5 dolls."

LEMON Learn

Learners Today. Leaders Tomorrow.

LEMONLearn

Learners Today. Leaders Tomorrow.

YOUR NEXT BOOK

Continue your child's learning with the next book in the LEAF Math Enrichment Program.

Toddler LEAF Math Workbooks 4, 5, 6

- Designed for early learners ages 2 and up.
- This book introduces numbers and number names up to 10, building on counting skills to strengthen number awareness through guided practice and repetition.
- A math games page at the end of each section helps parents naturally introduce math into everyday moments.
- A poster at the end of each section encourages your child's progress.

ISBN 978-1-971510-00-2

LEMON Learn

Learners Today. Leaders Tomorrow.

Made in the USA
Monee, IL
03 November 2026

67380774R00046